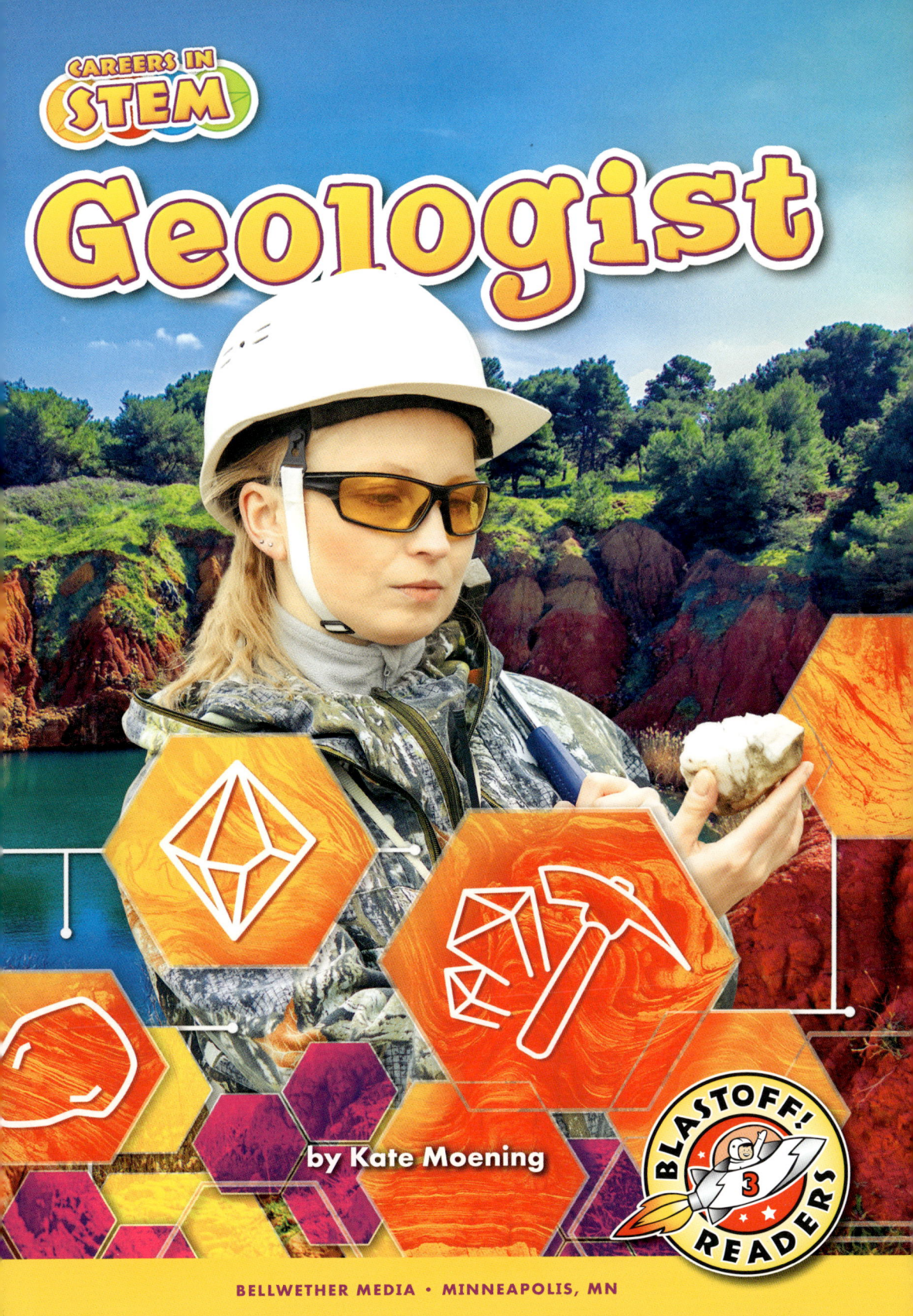

Geologist

CAREERS IN STEM

by Kate Moening

BELLWETHER MEDIA • MINNEAPOLIS, MN

Blastoff! Readers are carefully developed by literacy experts to build reading stamina and move students toward fluency by combining standards-based content with developmentally appropriate text.

 **Level 1** provides the most support through repetition of high-frequency words, light text, predictable sentence patterns, and strong visual support.

 **Level 2** offers early readers a bit more challenge through varied sentences, increased text load, and text-supportive special features.

 **Level 3** advances early-fluent readers toward fluency through increased text load, less reliance on photos, advancing concepts, longer sentences, and more complex special features.

★ **Blastoff! Universe**

This edition first published in 2023 by Bellwether Media, Inc.

No part of this publication may be reproduced in whole or in part without written permission of the publisher. For information regarding permission, write to Bellwether Media, Inc., Attention: Permissions Department, 6012 Blue Circle Drive, Minnetonka, MN 55343.

Library of Congress Cataloging-in-Publication Data

LC record for Geologist available at: https://lccn.loc.gov/2022005466

Text copyright © 2023 by Bellwether Media, Inc. BLASTOFF! READERS and associated logos are trademarks and/or registered trademarks of Bellwether Media, Inc.

Editor: Betsy Rathburn   Designer: Andrea Schneider

Printed in the United States of America, North Mankato, MN.

Table of Contents

Looking for Rocks … 4
What Is a Geologist? … 6
At Work … 10
Becoming a Geologist … 16
Glossary … 22
To Learn More … 23
Index … 24

Looking for Rocks

rock samples

A geologist gathers rock **samples**. She breaks pieces from a rock wall.

She writes down what she sees. She will test the samples in a **lab**!

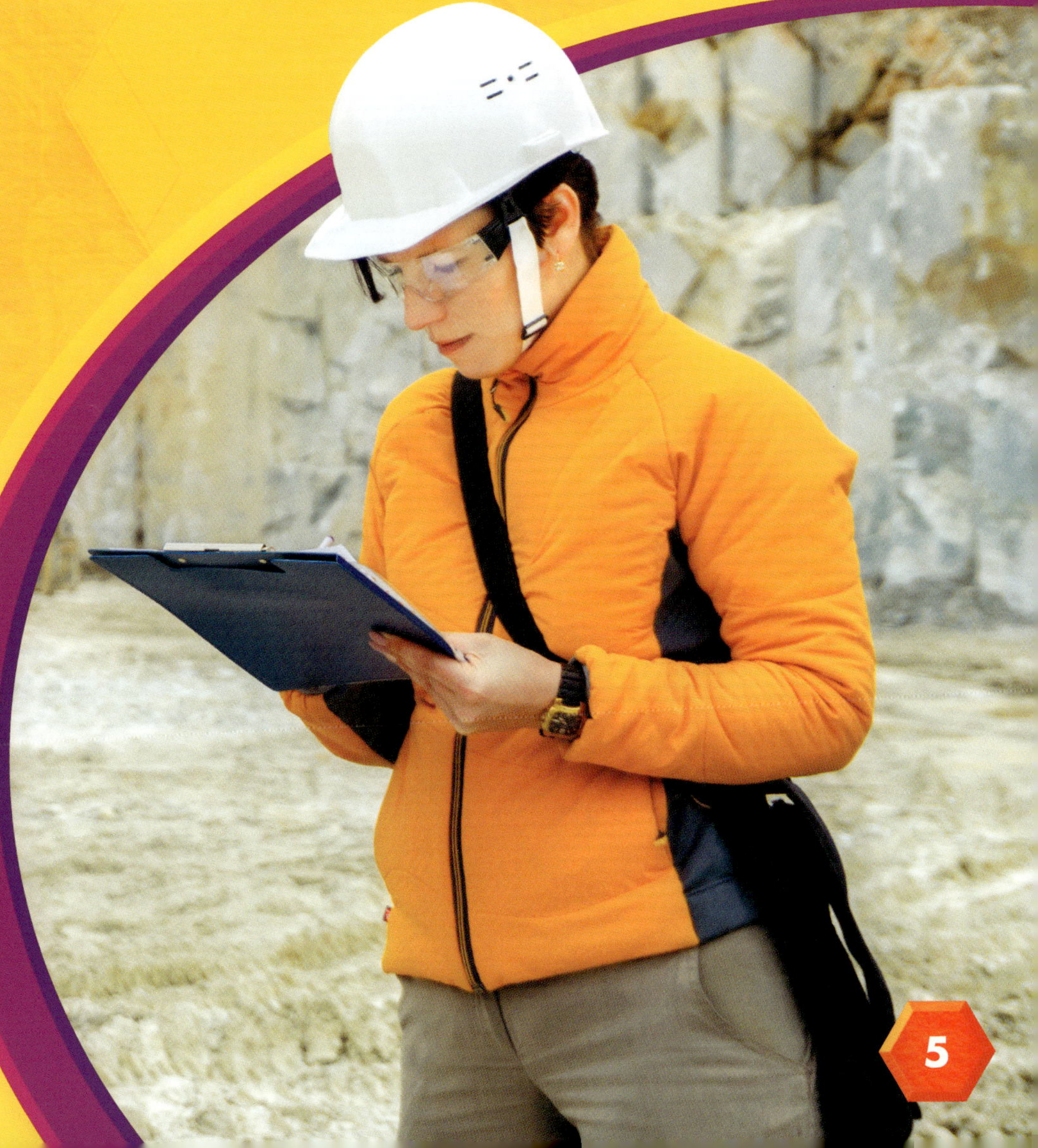

What Is a Geologist?

Geologists are scientists. They study Earth. They learn about Earth's history. They learn how rocks form.

Geologists collect samples. They do **surveys**. They go to mountains and oceans. Some work in space!

geologist collecting samples on the moon

Geologists test samples in labs. They study **minerals** in the samples. They find minerals that people can use.

Geologists also study events like **volcanic** blasts. This helps **predict** future events.

Famous Geologist

Name: Dawn Wright

Born: April 15, 1961

Birthplace: Baltimore, Maryland

Schooling: Wheaton College, Texas A&M University, University of California Santa Barbara

Known For: geologist who studies how the ocean floor forms and helped map the Pacific Ocean floor

At Work

Some geologists study rock history. They learn how rocks change.

Geologists test samples with computers. They learn when volcanoes blew up. They find the age of mountains!

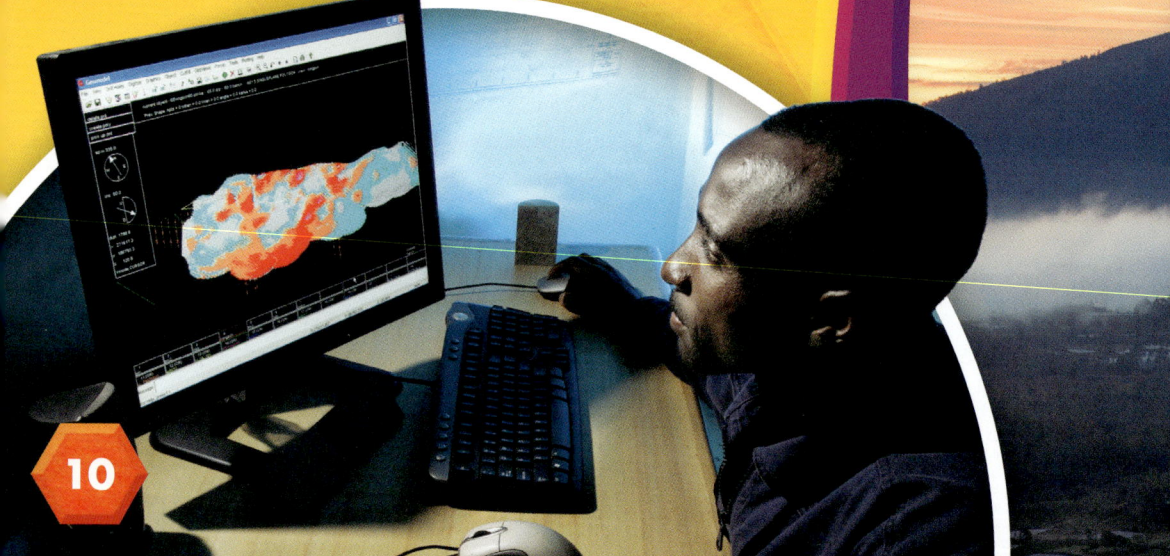

Geologists make maps of Earth's surface. They study how it forms.

Geologists also study how Earth changes over time. They learn about its past. They predict its future.

Geology in Real Life

safety plans for volcanic blasts

finding oil

products made with minerals

Surveys help geologists find different rocks. They look for rocks with important minerals inside.

Using STEM

Science — test rocks to learn about them

Technology — use computers to study rock history

Engineering — make sure land is safe to build on

Math — measure distances and angles to make maps of Earth's surface

geologist on a survey

Geologists remove the minerals. They go into things like computers and toothpaste!

Becoming a Geologist

Geologists work in all kinds of weather. They must be able to work outdoors.

Geologists read maps.
They study charts, too.
They must be good at math.

map

Many geologists study **earth science** in college. They also study math and computers. Some study **engineering**.

Some geologists continue school after college. They pick an area of geology to study further.

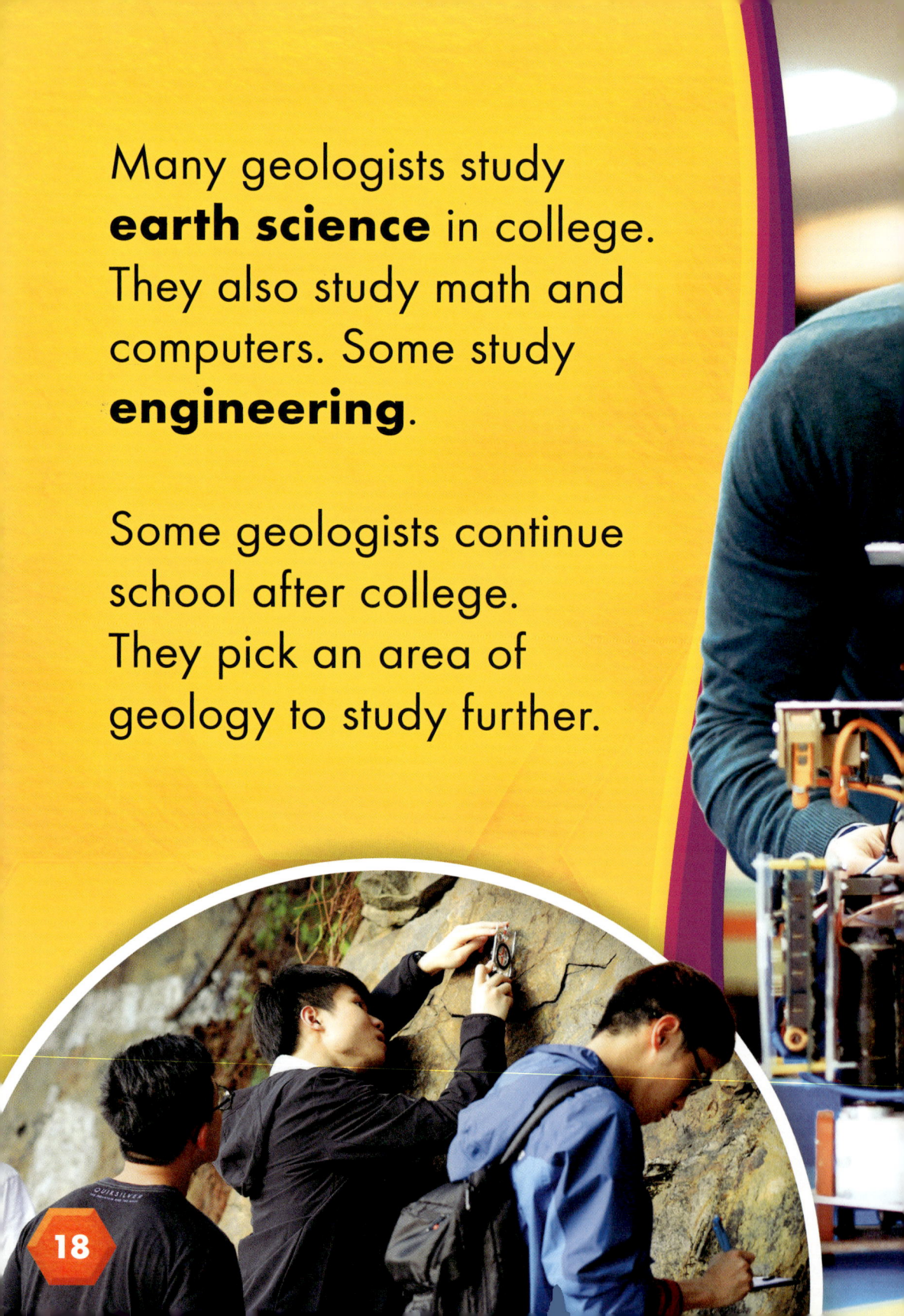

engineering students

Geologists sometimes do risky work. Most take a test to get a **license**. This shows that they can work safely.

How to Become a Geologist

1. study earth science and math in college

2. continue studying a special area of geology

3. take test to get a license

4. find a job at a business or lab

Geologists help us understand Earth!

Glossary

earth science—any of the sciences that deal with Earth or its parts; geology is an earth science.

engineering—the work of designing and building machines, systems, or structures

lab—a building with special tools to do science experiments and tests

license—an official document that gives geologists permission to work

minerals—naturally occurring substances found in rocks, sands, and soils

predict—to say that something will or might happen in the future

samples—small amounts of rocks that give information about the rocks' history or the area where they were taken from

surveys—acts of measuring and studying an area of land

volcanic—related to volcanoes; volcanoes are holes in the earth that let out hot ash, gas, and melted rock called lava.

To Learn More

AT THE LIBRARY

Huntington, Amy. *How to Make a Mountain*. San Francisco, Calif.: Chronicle Kids, 2022.

O'Mara, John. *How Do Rocks Form?* New York, N.Y.: Gareth Stevens Publishing, 2020.

Stroud, Jackie. *Under Your Feet*. New York, N.Y.: DK Publishing, 2020.

ON THE WEB

Factsurfer.com gives you a safe, fun way to find more information.

1. Go to www.factsurfer.com.
2. Enter "geologist" into the search box and click 🔍.
3. Select your book cover to see a list of related content.

Index

charts, 17
college, 18
computers, 10, 15, 18
Earth, 6, 12, 13, 21
earth science, 18
engineering, 18, 19
future, 9, 13
geology in real life, 13
history, 6, 10
how to become, 20
lab, 5, 8
license, 20
maps, 12, 17
math, 17, 18
minerals, 8, 14, 15
mountains, 6, 10

oceans, 6
predict, 9, 13
rocks, 4, 6, 10, 14
samples, 4, 5, 6, 8, 10
space, 6
surveys, 6, 14, 15
test, 5, 8, 10, 20
toothpaste, 15
using STEM, 14
volcanic blasts, 9, 10, 11
weather, 16
Wright, Dawn, 9

The images in this book are reproduced through the courtesy of: Evgeny Haritonov, front cover (geologist), pp. 4-5, 5, 20-21 (geologist); staraldo, front cover (background); Sebastian Janicki, pp. 3, 23 (quartz); Stephen Barnes, p. 4 (rock samples); Heritage Image Partnership Ltd/ Alamy, p. 6 (geologist collecting samples); Keith Douglas/ Alamy, pp. 6-7; Olaf Doering/ Alamy, pp. 8-9; Holly Mazour, p. 8 (mineral); Flickr, p. 9 (Dawn Wright); Greenshoots Communications/ Alamy, pp. 10 (inset), 16-17, 17; Erdem Summak, pp. 10-11; Chris Pearsall/ Alamy, pp. 12-13; Kuryanovich Tatsiana, p. 13 (safety plans for volcanic blasts); Evgeny_V, p. 13 (oil); Kenishirotie, p. 13 (products made with minerals); Science History Images/ Alamy, pp. 14-15; James Jiao, p. 18 (inset); NDAB Creativity, pp. 18-19; robertharding/ Alamy, pp. 20-21 (top); 3DMI, p. 22 (hammer); Albert Russ, p. 23 (agate).